AMAZING ANIMALS
OF THE WORLD

Greater Flamingo – Hooded Seal

Volume 12

GROLIER
EDUCATIONAL

Published by Grolier Educational Corporation

Remmel Nunn, *Vice President and Publisher*

Robert Hall, *Senior Vice President and National Sales Director*

Beverly A. Balaz, *Vice President, Marketing*

Laraine Balk, *Vice President, Operations*

Molly Stratton, *Editor, New Reference Titles*

Editorial development by the Grolier Reference Group

Lawrence T. Lorimer, *Vice President and Editorial Director*

Doris E. Lechner, *Managing Editor*

Bryan Bunch, Katia Lutz, Gisele Carruth, Sally Bunch, *Editors*

Douglas Lancaster, *Consulting Biologist*

Rebecca Lehmann-Sprouse, *Designer*

Bryan Bunch, *Guide Author*

Manufactured by Grolier Incorporated

Joseph J. Corlett, *Director of Manufacturing*

Christine L. Matta, *Senior Production Manager*

Pamela J. Terwilliger, *Production Manager*

Ann E. Geason, *Production Assistant*

Published by Grolier Educational Corporation 1995
© 1995 Grolier Inc.

Reprinted in 1998
Printed and bound in the United States of America

ISBN 0 – 7172 – 7396 – 2

Contents

How to Use This Book

Environment Symbols

Animals live where they do because the surroundings suit them. Some animals do not cope well with cold weather, so they live in areas where it is hot. Animals that get their food from trees likely live in or near forests. The natural surroundings in which an animal lives is called an environment. There are many different types of environments. Eight general environments describe the areas where animals live. A special environment symbol appears at the top of the page. This symbol tells you in which of the 8 environments the animal prefers to live. By remembering the 8 symbols, you can get a quick idea of any animal's environment.— *For more explanation see Kinds of Environments, page 46.*

Arctic and Antarctic

Deserts

Cities, Towns, and Forests

Forests and Mountains

Rain Forests

Oceans and Their Shores

Lakes, Rivers, and Their Shores

Grasslands and Savannas

Class Symbols

Scientists show how groups of living things are related by classifying them, a science called taxonomy. The classifications in which animals are placed range from the very general (kingdom and phylum) to the very specific (genus and species). Each phylum is made up of several classes. One of these symbols appears on the top of each page to identify the class to which the animal belongs. — *see How Animals are Classified, page 48.*

Arthropods

Fish

Mammals

Reptiles

Birds

Amphibians

Other, Invertebrates

Endangered/Extinct Symbols

These animals deserve special attention. Extinct animals no longer exist; endangered animals may soon become extinct. Special red symbols appear on the top of the animal pages next to the animals name to indicate endangered or extinct animals.

Endangered Animals

Extinct Animals

Maps

A map of a section of the world is featured for each animal. Within that section is a special shaded area. The shaded area is where the animal lives. For some, the maps show the whole world and are almost entirely shaded. For others, the shaded areas are very small. Since these animals live in very limited areas, there are probably very few of them. More information on where the animal lives and its characteristics appears above the map. An example of a typical map is shown below.

Home: Central Africa

Finding the Facts

symbol for endangered or extinct animal

environment symbol

animal name
scientific name

classification

animal data

map showing where animal lives

type of animal (color coded for easy sorting)

animal profile

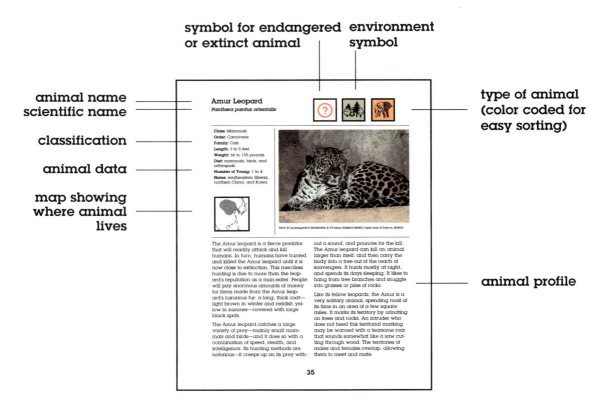

Greater Flamingo

Phoenicopterus ruber

Class: Birds
Order: Stilt-legged birds
Family: Flamingos
Height: to 5¼ feet
Weight: 5 to 9 pounds
Diet: algae, protists, and small invertebrates
Number of Eggs: 1
Home: Mediterranean region, Middle East, India, parts of Africa and Madagascar, and Caribbean region

PHOTO: Ardea, © ATP-Editions ROMBALDI MCMXCII, English version © Grolier Inc. MCMXCII

A group of flamingos in flight looks like a cloud colored pale pink or bright red. In the lagoons of the Yucatan peninsula of Mexico, flamingos are nearly red, while they are almost white in Europe. Their color depends on the foods they eat.

Flamingos nest in swampy areas in colonies of thousands of birds. They are good waders since, of all birds, flamingos have the longest neck and legs in proportion to their body. Their nest is established shortly before eggs are laid. It is a 15-inch-high cone of mud, stones, and shells. The single off-spring is nourished by food regurgitated by its parents. When they get older, the young flamingos assemble in large groups.

Flamingos are "filter feeders." Their curved beak has screens on the edges that filter solids from swamp water rich in tiny invertebrates, including crustaceans and mollusks. A flamingo searches for food by moving its bill upside down through the water. Only suitably-sized food passes through the screens. Sand and mud are thrown out by the pumping action of the flamingo's large tongue.

Ancient Romans used to enjoy the flamingo's tongue as a luxury food. People have also used flamingo feathers for decoration. Parks and racetracks are often graced with flamingos that have been encouraged to nest there to add color and beauty to these sites.

Greater Roadrunner
Geococcyx californianus

Class: Birds

Order: Cuckoos, touracos, and hoatzins

Family: Cuckoos

Length: 20 to 24 inches

Weight: 6 ounces

Diet: insects, reptiles, small birds, and fruit

Number of Eggs: 3 to 6

Home: southwestern United States and Mexico

PHOTO: Shaw/NHPA, © ATP-Editions ROMBALDI MCMXCII, English version © Grolier Inc. MCMXCII

Like the cartoon character that torments "Wile E. Coyote," real-life roadrunners would rather race along the desert sand than fly in the air. They do fly, however, using their short wings for short trips. A roadrunner pursuing a ground-dwelling insect or lizard can reach speeds of 15 miles per hour.

According to cowboy legends, greater roadrunners are clever enough to outwit rattlesnakes, and will go out of their way to pick a fight—stabbing a snake to death with lightning-quick jabs. The roadrunner's amazing speed allows it to escape predators by scooting between rocks and cacti. It is also one of the few wild animals fast enough to dodge a speeding car. Most important, the bird's agility enables it

to pounce on the insects, lizards, mice, and small snakes that make up its diet.

The only time of day in which the roadrunner is not quick on its feet is the early morning. This is because its body temperature drops considerably during the cold desert night. In the morning, it must bask in the sun until its temperature returns to normal.

Greater roadrunners mate for life. Together a mated pair will build a nest slightly off the ground—in low trees, shrubs, or cacti. The female roadrunner will lay three to six white or yellowish eggs that hatch in about three weeks. Both parents share in the care of the newborn chicks.

Greater Siren

Siren lacertina

Class: Amphibians

Order: Amphibians with long tails

Family: Sirens

Length: 20 to 40 inches

Diet: aquatic invertebrates and vegetation, insects, and small fish

Method of Reproduction: egg layer

Home: eastern coastal plain of the United States

PHOTO: © Tom McHugh/PHOTO RESEARCHERS, © ATP-Editions ROMBALDI MCMXCIII, English version © Grolier Inc. MCMXCIII

Like Peter Pan, the greater siren doesn't want to grow up—and it gets its wish. Although it certainly grows in size and weight as it ages, the greater siren stays in its immature larval state. This amphibian does not go through the kind of physical changes that tadpoles and other larval amphibians undergo on their way to becoming adult frogs, newts, and salamanders.

Sirens get their name from the mermaids of ancient mythology, who sang to sailors and, by distracting them, caused the men to crash onto rocks. It's hard to understand why anyone would give such a romantic name to this small water serpent. It is so slimy that most scientists put a sock over their hands before they'll even pick one up!

The greater siren has tiny forelegs with just four toes on each foot. It has no hind legs. Nevertheless, the siren can dig into the mud at the bottom of a stream. Sometimes it even uses its dinky forelegs to crawl onto dry land. When out of the water, the greater siren can sometimes be heard making a soft, squeaking sound.

During a drought the siren has an ingenious way of surviving. Just before its watery home totally evaporates, the siren digs into the pond bottom and builds itself a cocoon of mud. Within this case the siren seals its body with slimy skin secretions. In this way, it keeps from dehydrating.

Green Anole

Anolis carolinensis

Class: Reptiles

Order: Lizards

Family: Iguanas

Length of the Body: 2½ in.

Length of the Tail: 3 to 4 in.

Weight: 2 to 3 pounds

Diet: insects and spiders

Number of Eggs: 1 or 2

Home: southeastern United States to Texas; Cuba and Honolulu

PHOTO: Lanceau/NATURE, © ATP-Editions ROMBALDI MCMXCII, English version © Grolier Inc. MCMXCII

Most people know the green anole as the attractive little "chameleon" sold in pet stores. It is the only species native to the continental United States. Chameleons, of course, are famous for their ability to change color. They do so when frightened, angry, or otherwise excited. They also change shades to match their surroundings. The green anole can switch colors quite rapidly. But its palette is limited to greens and browns. This species can also be recognized by its pink throat fan, a loose flap of skin beneath its chin.

Male green anoles engage in dramatic fights during mating season. If two males meet, they both turn from brown to bright green in excitement. Then they begin shaking their heads and flapping their pink throat fans. If one male does not back down at this point, the lizards start darting from side to side, taking swipes at one another. Eventually one of the males admits defeat by slinking away, wearing a dull shade of yellowish brown. The winner turns an even brighter shade of green.

The green anole makes a friendly and lively pet. But caring for it requires a lot of responsibility and effort. This reptile will not drink water from a bowl. You must quench its thirst with droplets sprinkled on fresh leaves. It also needs live insects and spiders for food. In return for your care, this pet will learn to stay on your shoulder and even eat flies and worms right out of your hand!

Green Frog
Rana clamitans

Class: Amphibians
Order: Frogs and toads
Family: True frogs
Length: 2¼ to 4 inches
Diet: insects
Method of Reproduction: egg layer
Home: eastern United States and the Maritime provinces of Canada

PHOTO: Shaw/NHPA, © ATP-Editions ROMBALDI MCMXCII, English version © Grolier Inc. MCMXCII

The green frog is a familiar animal to many North Americans. It lives in colonies along the shores of small, shallow ponds and streams. Like other frogs and toads, the green frog undergoes a remarkable change of appearance as it grows, a process called metamorphosis.

Green frogs lay their eggs (spawn) in fresh water in the spring or summer. The eggs are not guarded; instead, the frog surrounds the eggs with a jellylike material. After a few days, the eggs hatch, and larvae emerge. The larvae are called tadpoles. Much like fish, the tadpoles breathe through gills. At first the gills are on the outside of the body, but they soon become covered with a fold of skin. Eventually lungs grow,

legs develop, the mouth widens, and the tail begins to disappear. By the time the tail is completely gone, the frog is ready for life on land. As an adult the frog will return to that very same body of water to spawn.

Male green frogs defend their territory by standing on their hind legs and hugging each other. They call mates by making a loud noise, or croak, that sounds like a loose banjo string. Green frogs are equipped with a long, sticky tongue that is attached to the front of their mouth. When an insect flies by, the frog simply flicks out its tongue and makes the kill. In this way, green frogs catch and eat an enormous amount of insects.

10

Green Iguana

Iguana iguana

Class: Reptiles

Order: Lizards and snakes

Family: Iguanas and their relatives

Length: up to 6½ feet

Diet: plant matter and insects

Number of Eggs: 20 to 70

Home: Central America and South America

PHOTO: Gohier/NATURE, © ATP-Editions ROMBALDI MCMXCII, English version © Grolier Inc. MCMXCII

In a fight between a green iguana and a dog, the dog often loses. This is because of the iguana's not-so-secret weapon: a strong, muscular tail that it whips at its attacker. A well-placed strike of the tail causes enough pain to convince the attacker to look elsewhere for a meal. The iguana may also try to bite its predator with its sharp teeth and powerful jaws. Unfortunately, such weapons do little to protect green iguanas against people who hunt and eat iguanas and their eggs.

Running the length of the green iguana's back is a crest of large scales. On a male the crest may reach up to 3 inches. The green iguana also has a fold of loose skin, called a dewlap, hanging under its neck. Like the crest, the dewlap is larger in the male than in the female.

The green iguana prefers to live in treetops in tropical and subtropical forests. Its well-developed claws are useful in climbing. It often seeks to live near water, where it can take advantage of its excellent swimming ability to escape enemies. When threatened, the green iguana dives below the surface and swims underwater to a nearby bank. Some green iguanas live in shrubby coastal regions, where there is a dry season and a rainy season. During the rainy season, plants grow rapidly, and food is plentiful. The iguana stores large quantities of fat in its body. This fat helps it survive through the dry season, when food is scarcer.

Green Mamba
Dendroaspis angusticeps

Class: Reptiles
Order: Lizards and snakes
Family: Cobras, mambas, and coral snakes
Length: up to 6½ feet
Diet: mainly lizards and birds
Number of Eggs: 10 to 15
Home: eastern Africa, from Kenya south to Zimbabwe

PHOTO: © Tom McHugh/PHOTO RESEARCHERS, © ATP-Editions ROMBALDI MCMXCII, English version © Grolier Inc. MCMXCII

The green mamba spends most of its life in trees, where its leaflike coloration helps it blend in with the jungle background. This creature can move extraordinarily fast for a snake: one green mamba has been clocked speeding along at a remarkable 7 miles per hour—a rate at least twice that of any North American snake.

Like all mambas, the green variety packs an extremely poisonous venom. But when the green mamba sets out looking for its next meal, it moves about slowly, head held high with mouth open, exposing fierce-looking fangs. When the prey is within striking range, the mamba attacks. The fangs pierce through the prey, all the time injecting a whitish venom. The venom affects the victim's nervous system, especially that part that controls the heart and breathing.

The green mamba tends to avoid humans, and is, in fact, much less aggressive than other mamba species. Nonetheless, many humans are bitten each year by the green mamba, usually only after directly confronting or pursuing the snake. A person bitten by a green mamba faces a true medical emergency. The green mamba usually strikes at a human's legs or torso, and the venom works very rapidly. If the victim is not quickly treated with a special antivenin (a medication that neutralizes the venom), death is almost certain.

Green Moray

Gymnothorax funebris

Class: Bony fishes

Order: Eels

Family: Moray eels

Length: up to 8 feet

Diet: fish, octopuses, and other invertebrates

Method of Reproduction: egg layer

Home: tropical and subtropical waters off the east coasts of Central and South America

PHOTO: © Gregory Ochocki/PHOTO RESEARCHERS, © ATP-Editions ROMBALDI MCMXCIII, English version © Grolier Inc. MCMXCIII

The green moray is an eel that lurks among the holes and crevices of coral reefs. With its tail anchored around coral or rock, it reaches out its toothy mouth to snatch up passing octopuses. The green moray seldom bothers scuba divers unless harassed. Its bite is not poisonous, but the wound often becomes painfully infected.

Even when it is resting, the green moray presents a frightening face. It continually opens and closes its mouth as if to bite. Actually, it is only breathing. The moray gulps water into its mouth and then expels the water through its gills.

Despite appearances, the green moray's body is actually blue. It looks green because of a yellow slime over its skin. This slime, or mucus, protects the moray from scrapes on the sharp coral. The mucus also contains a natural antibiotic to help prevent infection.

Morays do not bother with courtship. Males and females simply release their eggs and sperm into the water, where the currents help them to mix. When a moray larva hatches, it looks like a long tree leaf. This kind of young is called "leptocephalus," meaning "flat-headed" in Latin. Leptocephalus larvae drift with the ocean currents, absorbing dissolved food directly into their body from the water. As they mature, the larvae grow teeth and begin to look like adult eels.

Green Sawfly

Rhogogaster viridis

Phylum: Arthropods

Class: Insects

Order: Wasps, ants, and bees

Family: True sawflies

Length: ³/₈ to ¹/₂ inch

Diet of the Larva: leaves of trees and herbs

Method of Reproduction: egg layer

Home: North America, Europe, and Asia

The sawfly is named for the sawlike tool at the end of the female's body. She uses it to cut into plant tissue to make small pockets for her eggs. The mother sawfly is choosy about the plants she lays her eggs on. Among her favorites are leafy willows, poplars, buttercups, and alders.

The newly hatched grubs of the green sawfly resemble tiny caterpillars. Each has three pairs of legs at the front of its body, no legs in the middle, and more than five pairs of legs in the back. Crawling like an inchworm, the grub steps down with its front feet, and then bends the middle of its body to move up its back feet. Sawfly grubs have a tremendous appetite for leaves and often gather in groups to eat this tasty food.

Like many insects the green sawfly changes its body shape several times before maturing into a winged adult. The adult has a thick waist, quite unlike the thin "threadwaists" of its relatives the bees, wasps, and ants. The adult lives for only one or two weeks. Within this time, it may never feed. At most, it may sip a drop of nectar with some pollen. During its brief life, the adult green sawfly gathers in small swarms along the green edges of rivers and streams, in open woodlands, and over undisturbed meadows. There the sawflies mate and lay their eggs for the next year's generation of insects.

Green Sea Urchin

Strongylocentrotus droebachiensis

Phylum: Echinoderms

Class: Sea urchins and sand dollars

Order: Sea urchins

Family: Strongylocentrotidae

Diameter: about 3 inches

Weight: about 8 ounces

Diet: algae, barnacles, and dead animal matter

Method of Reproduction: egg layer

Home: North Atlantic, North Pacific, and Arctic oceans

PHOTO: © F. Stuart Westmorland/TOM STACK & ASSOCIATES, © ATP-Editions ROMBALDI MCMXCII, English version © Grolier Inc. MCMXCII

Despite its elaborate Latin name (try saying it three times fast), the green sea urchin is one of the most common creatures in the sea. It is most abundant in the cold waters of the northern oceans, where it lives in thick groups clustered on rocks. Green sea urchins reproduce so fast that they crowd out other types of urchins. Lobster fishermen pulling up their traps often find them filled with these prickly green balls instead of tasty lobsters and crabs.

The sea urchin's spines discourage predators from taking a bite. These quills have other important purposes as well. To begin with, all of the sea urchin's spines are attached to joints and muscles. By moving its spines, the sea urchin can bury itself in the sand or crawl short distances across the ocean floor. Many of the urchin's quills are not spines at all. Some are tubelike feet with suckers at their tips. The sea urchin uses these suckers to firmly attach itself to a rock or other solid surface. Other quill-like structures, called pedicellariae, have tiny, sharp jaws at their tips. The sea urchin uses them to nip at its enemies. If you turn a sea urchin over, you may be able to see what looks like a belly button. Actually, it is the sea urchin's chewing apparatus, which is called an Aristotle's lantern. It contains five sharp teeth that grind up food into digestible bits.

Green Turtle
Chelonia mydas

Class: Reptiles

Order: Turtles and tortoises

Family: Sea turtles

Length: up to 4 feet

Weight: commonly 250 to 300 pounds; but it can weigh up to 600 pounds

Diet: algae (adult); small animals and plants (young)

Number of Eggs: 65 to 200

Home: tropical and subtropical oceans worldwide

PHOTO: Lanceau/NATURE, © ATP-Editions ROMBALDI MCMXCII, English version © Grolier Inc. MCMXCII

The green turtle doesn't get its name from the color of its shell, but the the greenish tint of its fat. Since ancient times, its meat has been considered a luxury food, and its eggs are eaten in great quantity in parts of the world. Today, the green turtle is in danger of extinction, although it is protected in a number of countries. Its numbers are decreasing because it has been widely hunted and its mating places are being destroyed.

The green turtle lays up to 200 eggs at any time of the year, but usually in summer. Many of its mating grounds are on the beaches of the Caribbean Sea. While it lives alone most of the year, large groups gather to lay their eggs on these beaches. Females leave the water and slowly climb up the beach until they are above the high-tide line. There, they dig a hole about 18 inches deep, lay their eggs, and cover them with sand. After 6 or 7 weeks, the young turtles hatch, and try to get to the sea. But many enemies (birds, dogs, snakes, and lizards) wait to attack and eat them. Very few turtles will reach the water. Young turtles that survive begin a journey into the open sea. Scientists study this journey by putting transmitters on the animals that send signals to satellites, which are then transmitted back to earth. A large green turtle farm in the Cayman Islands releases thousands of turtles into the Caribbean every year. The young turtles are raised in captivity and returned to the ocean when they are older and more likely to survive.

Green-backed Heron

Butorides striatus

Class: Birds
Order: Stilt-legged birds
Family: Herons and bitterns
Length: 15 to 19 inches
Wingspan: 20 to 24 inches
Diet: mainly fish
Number of Eggs: 2 to 4
Home: South America, Africa, Asia, Polynesia, and Australia

PHOTO: Grospas/NATURE, © ATP-Editions ROMBALDI MCMXCII, English version © Grolier Inc. MCMXCII

Night has fallen over the mangrove forest—a dense growth of trees and shrubs common along tropical rivers and seashores. At the edge of the water lurks a green-backed heron. Suddenly the heron lunges into the water and spears a fish with its bill. Another heron flying overhead also spots food and dives into the water. This scene is repeated over and over again as these birds search for food. The green-backed heron always looks hungry. Indeed, when the food supply is plentiful, it is a greedy eater. Mainly fish, but also frogs, crabs, mollusks, insects, and worms make up its diet.

The green-backed heron is a thin bird with long legs, a long neck, and a pointed bill. It generally remains in the same area all year, though some may migrate when the rains stop and the ground dries up. In flight a heron usually folds its neck so that its head is tucked back to the shoulders. The legs stretch backward, trailing behind the body. An excellent flyer, it moves easily and surely among the mangrove trees. Both parents incubate the eggs and care for the babies in a nest made from twigs.

In some places the green-backed heron is a shy, secretive creature. During the day, it rests in tall grass and under overhanging branches, flying at sundown to look for food. Herons that live near city ponds and mud banks are accustomed to people and will often feed during the day.

Green-winged Teal

Anas crecca

Class: Birds

Order: Waterfowl

Family: Geese and ducks

Length: 13 to 16 inches

Wingspan: 20 to 25 inches

Diet: seeds; small aquatic mollusks, insect larvae, and other animals

Number of Eggs: usually 8 or 9

Home: throughout most of the Northern Hemisphere

PHOTO: Sauer/NATURE, © ATP-Editions ROMBALDI MCMXCIII, English version © Grolier Inc. MCMXCIII

The green-winged teal is a dabbler. It usually feeds in shallow water, sitting on the water's surface and dipping ("dabbling") its bill and head into the water in search of food. The teal has a broad, flat bill with grooves at the sides. These grooves form a strainer that removes food from the water.

The young green-winged teal feeds mainly on insect larvae and tiny crustaceans. As it grows older, the teal adds seeds and plants to its diet. Adult green-winged teal eat mainly seeds of water plants such as sedges and pondweeds. In autumn, they sometimes visit fields to eat corn and other grain left behind after harvesting. They may also eat berries, wild grapes, and acorns.

Green-winged teal feed and travel in groups, or flocks. They migrate northward to their breeding grounds in early spring. There they build nests on the ground, usually near the edge of lakes or ponds. Only the female incubates the eggs and cares for the young. Young teal can swim soon after hatching, and are able to fly when they are about seven weeks old. In autumn the teal migrate southward to warmer lands.

These teal are named for a very obvious physical characteristic—the green patches on their wings. For most of the year, the male has a bright-brown head and a white stripe above each wing. In midsummer, he sheds this coat and grows a coat that resembles the speckled brown coat of the female.

Greenhouse Millipede

Oxidus gracilus

Class: Millipedes

Order: Centipedes

Famly: Strongylosomid milli-pedes

Length: 1 inch

Number of Legs: 72

Diet: decaying plant matter

Number of Eggs: up to 300

Home: greenhouses world-wide

PHOTO: Allistar Shay/ANIMALS ANIMALS, © ATP-Editions ROMBALDI MCMXCII, English version © Grolier Inc. MCMXCII

Greenhouse millipedes originally inhabited tropical Asia and Africa. But long ago a few of them found their way into someone's garden green-house. There they thrived in the con-stant warm and moist human-made climate. The greenhouse millipede was able to travel around the world by hitching rides on plants or in bags of soil that were traded or sold between greenhouses. Today this millipede is a familiar sight to greenhouse workers. Fortunately, it rarely eats living plants. Sometimes it eats the roots of young plants, but rarely enough to hurt them.

The name millipede means "a thou-sand legs," just as the name "cen-tipede" means "a hundred legs." Neither figure is accurate. Millipedes have, at most, just over a hundred legs. A more important difference between millipedes and centipedes is the num-ber of legs on each of the animals' body segments. Adult millipedes have two pairs of legs on each of their body segments, while centipedes have only one pair.

Millipedes are simple animals. They have no sense of smell, but can taste sugar. They continually tap the ground with their antennae, searching for any-thing sweet. Some millipedes have primitive eyes, but greenhouse milli-pedes have none. They can, however, sense light through their skin! If you shine light on them, they will scurry around until they find a dark place to hide.

Grunion

Leuresthes tenuis

Class: Bony fishes

Order: Silversides, flying fishes, and toothed carps

Family: Silversides

Length: 5 to 7½ inches

Diet: small invertebrates

Method of Reproduction: egg layer

Home: coastal waters off California and Baja California

PHOTO: Foott/OKAPIA, © ATP-Editions ROMBALDI MCMXCIII, English version © Grolier Inc. MCMXCIII

Grunions are best known for their unusual spawning behavior. These fish spawn, or produce eggs, in the moonlight on spring and summer nights when tides are exceptionally high. Soon after the highest wave of the night, the grunions let a smaller wave carry them onto a sandy beach. Thousands of grunions may come ashore at one time.

The female grunion quickly burrows tailfirst into the wet sand. When the back half of her body is buried, she lays 1,000 to 3,000 tiny pink eggs. The male wraps his body around the upper, exposed part of the female and releases sperm, which seeps down through the sand to fertilize the eggs. The fish then wiggle back to the water and are carried out to sea by the next wave. The entire process takes less than a minute.

The eggs remain buried in the sand until 15 days later, when the tides are again exceptionally high. The high waves wash the eggs out of the sand. Within minutes the eggs hatch, and the young grunions are carried out to sea. Grunions are about ¼ inch long at birth. They grow quickly, reaching about 5 inches when they are a year old. By then they are ready to return to the sandy beach to lay eggs of their own. If grunions are not caught by people or killed by marine predators or disease, they live for about three to four years. People are allowed to catch spawning grunion by hand at certain times of the year.

20

Guinea Pig

Cavia aperea porcellus

Class: Mammals

Order: Rodents

Family: Cavies

Length: about 10 inches

Weight: 1 to 2 pounds

Diet: grasses, hay, leaves, bark, fruits, roots, blossoms, and seeds

Number of Young: 1 to 4

Home: worldwide

PHOTO: © Robert Maier/ANIMALS ANIMALS, © ATP-Editions ROMBALDI MCMXCIV, English version © Grolier Inc. MCMXCIV

The guinea pig, a popular and gentle children's pet, was domesticated 3,000 to 6,000 years ago. Villagers in the mountains of Peru tamed the guinea pig's ancestor, the wild cavy and kept it around their huts. These early guinea pigs, fattened on garbage, were a favorite meal of the ancient Peruvians. Today they are still raised as farm animals in various parts of Central and South America.

In the 16th century, traders brought a cargo of domestic guinea pigs to Europe, and they immediately became popular pets. The guinea pig shares many traits with the wild cavy, but it is considerably calmer and gentler around humans. And unlike its wild ancestor, the domestic guinea pig purrs like a cat. Adult males purr louder and more often than females and piglets.

Despite its long, sharp teeth, a guinea pig will rarely bite its owner. It is easy to care for and thrives on such common foods as oat flakes, apples, lettuce, and carrots. Pet guinea pigs also breed easily and mate throughout the year. Newborn piglets emerge fully developed, with open eyes and ears. Guinea pigs may live to seven years of age.

In addition to making fine pets, guinea pigs are important laboratory animals. They share many anatomical similarities with humans, and so are used in various types of medical and drug research.

Guppy
Poecilia reticulata

Class: Fish

Order: Atheriniformes

Family: Live-bearing topminnows

Length: up to 2 inches

Diet: aquatic insects, algae, and fish eggs

Number of Eggs: 10 to 100

Home: native to the West Indies and northern South America; introduced elsewhere

PHOTO: © Peter Gathercale/ANIMALS ANIMALS, © ATP-Editions ROMBALDI MCMXCII, English version © Grolier Inc. MCMXCII

The guppy is one of the best-known aquarium fish in the world. It is an easy fish to keep because it doesn't mind small variations in water temperature. The guppy matures rapidly and produces large quantities of eggs. For these reasons, guppies are used in science laboratories for experiments in genetics, fish physiology, and breeding behavior.

Guppies have also served humans in other ways. They have been introduced around the world in places such as Argentina, Hawaii, and Singapore to help control malaria outbreaks. How can guppies fight a human disease? They eat the floating larvae of the mosquitoes that carry the malaria germ.

By reducing the population of malaria mosquitoes, the guppies have saved thousands of human lives.

Guppies are viviparous, which means they give birth to live young instead of eggs. The first thing a newly born guppy will do is swim to the surface to gulp air to fill its swim bladder. This helps the baby guppy swim—and it had better swim fast.

Guppy mothers will eat their own young if given a chance. For this reason, fish breeders often put their pregnant guppies in "maternity cages." The cages have holes big enough to let the babies escape, but small enough to keep their mothers at bay.

Gypsy Moth
Porthetria dispar

Phylum: Arthropods

Class: Insects

Order: Butterflies and moths

Family: Tussock moths and their relatives

Wingspan: 2½ inches (female) 1 inch (male)

Diet: tree leaves

Method of Reproduction: egg layer

Home: native to Europe and Asia; introduced into North America

PHOTO: © D. Wilder/TOM STACK & ASSOCIATES, © ATP-Editions ROMBALDI MCMXCIII, English version © Grolier Inc. MCMXCIII

Leopold Trouvelot thought he had a wonderful idea when he decided to start a silk industry in the United States. He imported some silk-producing caterpillars from Europe, but soon abandoned the project. Unfortunately, by 1868 some of these gypsy moth caterpillars escaped into woods in Massachusetts. Lacking natural enemies, the gypsy moths quickly expanded their range and became serious pests on trees. Trouvelot's "wonderful idea" turned out to be a disaster.

Female gypsy moths are so heavy that they can barely fly. The smaller males fly about looking for the females, who give off a chemical odor that attracts the males. After mating, a female lays a mass of up to 1,000 eggs. The moths then die, after living less than two weeks as adults. The eggs hatch the following spring. The tiny caterpillars are the gypsies that give this insect its common name. They climb to the tops of trees and drop off on silken threads produced by glands in their abdomen. Wind carries them to a new location. Then they begin eating tree leaves. Gypsy-moth caterpillars can devastate acres of trees. At times during summer, the caterpillars eat so much foliage that the trees are totally bare and look much like they do in the dead of winter. In late June the caterpillars are about 2 inches long. They stop eating, and each caterpillar spins a cocoon around itself. During the next two weeks it metamorphoses, changing into an adult moth.

Haddock

Melanogrammus aeglefinus

Class: Bony fishes

Order: Cod and their relatives

Family: Cod

Length: usually up to 2½ feet

Weight: up to 10 pounds (record: 37 pounds)

Diet: mainly bottom-dwelling invertebrates

Method of Reproduction: egg layer

Home: North Atlantic Ocean

PHOTO: © P. Morris/ARDEA LONDON LTD., © ATP-Editions ROMBALDI MCMXCIII, English version © Grolier Inc. MCMXCIII

Haddock live on the ocean floor, usually in water 100 to 500 feet deep, seldom rising more than a few feet from the bottom. They feed on almost any kind of invertebrate that inhabits this part of the ocean, including starfish, small worms, crabs, sand dollars, sea cucumbers, and mollusks. Haddock also prey on small fish, squid, and shrimp that swim near the bottom.

Haddock look like cod, but they are smaller and do not have the cod's characteristic speckled sides. Haddock are easy to recognize by the black blotch behind their gills. In springtime, haddock migrate to their spawning grounds. A medium-sized female may lay more than 150,000 eggs. The young haddock live near the ocean surface for a while, but change to bottom dwellers before they are a year old. Many eggs and young do not survive because they are eaten by fish and other sea life.

Like cod, haddock is an important food fish. Sometimes it is smoked, a practice that began in the 1700s in Findon, Scotland. Today smoked haddock is known as finnan haddie—an abbreviation of Findon haddock.

People have overfished haddock; that is, they have caught haddock faster than the fish have been able to reproduce. As a result, haddock populations have fallen, and fishermen find it more and more difficult to locate large specimens.

Hadrosaurus

Hadrosaurus

Class: Reptiles

Order: Bird-hipped dinosaurs

Family: Duck-billed dinosaurs

Length: about 30 feet

Diet: ferns, horsetails, leaves, twigs, and blossoms

Method of Reproduction: egg layer

Home: North America

Date of Extinction: about 65 million years ago

PHOTO: American Museum of Natural History, © ATP-Editions ROMBALDI MCMXCIV, English version © Grolier Inc. MCMXCIV

Hadrosaurus, a name that means "giant lizard," was the first dinosaur ever discovered in North America. Its bones were unearthed in New Mexico in 1858. Soon after, the skeleton was reconstructed by Joseph Leidy at the University of Pennsylvania. Leidy knew enough about anatomy to realize that this dragonlike dinosaur reared up on its back legs to run. *Hadrosaurus* did not stand completely upright. Rather, it tilted back just enough to lift its small front legs off the ground. As it ran, the hadrosaur used its massive tail to balance the weight of its chest and head.

Left undisturbed, hadrosaurs probably spent their time on all fours, grazing in the ancient woodlands of North America. However, this dinosaur was a favorite prey of *Tyrannosaurus*, the largest meat-eating dinosaur of them all. Hadrosaurs, which lived in herds, must have always been on the lookout for their horrible enemy. Given any warning, a hadrosaur running on two legs could have easily outraced a tyrannosaur.

Hadrosaurus was a duck-billed dinosaur. Duckbills, a tremendously successful family of plant eaters, were among the last dinosaurs to survive on Earth. They all had a broad, flat snout with a toothless beak. The back of the duckbill's powerful jaws were filled with hundreds of molars, enabling it to chew the toughest of plants.

Hairy Doris

Acanthodoris pilosa

Phylum: Mollusks

Class: Snails and slugs

Order: Nudibranchs

Family: Doridacean nudibranchs

Length: 1¼ inches

Width: ⅝ inch

Diet: bryozoans

Method of Reproduction: egg layer

Home: coastal waters of northern Europe and eastern North America

PHOTO: © David Nance/ARDEA, © ATP-Editions ROMBALDI MCMXCIII, English version © Grolier Inc. MCMXCIII

The hairy doris is an ocean-dwelling creature that lives hidden among seaweeds in shallow seawater. It glides slowly and gracefully, with wavy movements of its body. On its head are two eyes and two tentacles. The tentacles function like a person's nose; they pick up smells from the surrounding water. As the hairy doris moves along, it searches for a certain kind of prey: an animal called the porcupine bryozoan. Bryozoans are tiny invertebrates that live together in large colonies. The colonies have various shapes, depending on the species of bryozoan. When the hairy doris finds porcupine bryozoans, it uses its rasp-like tongue to scrape off pieces of the colony.

The hairy doris is a nudibranch. Nudibranchs are mollusks that do not have shells as adults and are shaped like slugs. Some people call them "sea slugs." Nudibranchs are often very beautiful. The hairy doris is whitish or purplish brown, and its back is covered with soft projections called tubercles.

This creature lays eggs in a gelatinous mass. It attaches one end of the mass to a solid object in the water, and then wraps the rest around the object. The larvae that hatch from the eggs swim about for a time before they settle to the bottom. The larvae have shells, but these shells soon disappear. The animals change, or metamorphose, into juveniles that look like miniature adults.

Hairy Woodpecker

Picoides villosus

Class: Birds

Order: Woodpeckers, toucans, and honeyguides

Family: Woodpeckers

Length: about 9½ inches

Diet: insects, nuts, and sap

Number of Eggs: usually 4

Home: Canada, United States, Mexico, and Central America

PHOTO: © Alvin E. Staffan/PHOTO RESEARCHERS, © ATP-Editions ROMBALDI MCMXCIV, English version © Grolier Inc. MCMXCIV

The courtship of the hairy woodpecker is long and a little odd. Rather than wait for spring, as do most animals, these birds begin courting in winter. Hairy woodpeckers also reverse the usual roles of male and female. It is the female who actively pursues the male. First she tries to catch his attention by drumming loudly on a tree trunk. Then she performs a courtship dance by quivering and fluttering her wings. The male performs a traditionally female task: he selects the nest site. Both birds share in the hard work of drilling a large nest hole in the trunk of a tree.

Once the hairy woodpecker lays her eggs, she and her mate take turns warming them. Their chicks hatch in about two weeks. Both parents feed the chicks, but each plays a slightly different role. The father woodpecker flies far from the nest. He makes few return visits, but brings plenty of large insects when he does. The mother stays closer to home. As she hunts for small insects for her babies, she listens for any sound of danger. If an enemy such as a cat approaches the nest, the mother tries to chase it away or distract it.

The hairy woodpecker catches insects that have burrowed inside the bark of trees. To reach them the bird uses its long, sticky tongue. There are tiny, sharp barbs at the tongue's tip. When extended, the bird's tongue reaches several inches.

Hairy-legged Vampire Bat
Diphylla ecaudata

Class: Mammals
Order: Bats
Family: Vampire bats
Length: 2½ to 3½ inches
Weight: about 1 ounce
Diet: blood of chickens and other birds
Number of Young: 1
Home: Central and South America

PHOTO: © Richard K. LaVal, © ATP-Editions ROMBALDI MCMXCII, English version © Grolier Inc. MCMXCII

The hairy-legged vampire bat feeds on the blood of chickens and other birds. It is an agile creature that approaches its prey by "walking" on all fours. Using the large thumbs on the ends of its wings as extra pairs of feet, it can run, hop, and jump along the ground, searching for its next victim.

The hairy-legged vampire bat hunts at night. It uses several methods to locate a victim. Excellent eyesight and a good sense of smell allow the creature to track down prey with ease. In addition the bat has heat-sensitive pits on its face that detect body heat given off by an animal. And the bat echolocates—that is, it sends out high-pitched signals that bounce off objects in its path. Echolocation works much the way sonar does on a submarine.

In preparation for a meal, the bat usually bites the legs of its prey. It makes a small cut in the skin, then laps up blood until it has satisfied its hunger. In just 10 minutes, the bat can drink its weight in blood! The main danger to the chicken is not the loss of blood, but the possibility of infection. Vampire bats often carry diseases that can kill birds.

Hairy-legged vampire bats are shy creatures that live in a variety of habitats, including dry forests, rain forests, grasslands, and farmlands. They roost in caves, hollow trees, abandoned mines, and houses. Contrary to popular belief, the hairy-legged vampire bat poses no threat to humans.

Hamadryas Baboon
Papio hamadryas

Class: Mammals

Order: Primates

Family: Old World monkeys

Length: 24 to 30 inches

Diet: plants, small animals, and honey

Home: the horn of Africa and the Arabian peninsula

PHOTO: Samba/NATURE, © ATP-Editions ROMBALDI MCMXCII, English version © Grolier Inc. MCMXCII

In ancient times, the hamadryas baboon was common in Egypt. People then thought it was the living form of the sun god Toth and the animal was often shown in paintings and statues. When they died, theses baboons were wrapped like mummies and placed in tombs.

In the past, hamadryas baboons formed very large troops that often damaged crops. In the late 1800s, however, the hamadryas was nearly killed off by European hunters. Today there are none left in Egypt, but some can still be seen in the horn of Africa and the Arabian peninsula. Hamadryas baboons live in family groups made up of 1 adult male and 1 to 9 females with their young. At night, theses small groups get together as a troop to sleep. As many as 700 baboons may rest on 1 cliff.

In the morning, the small groups of hamadryas separate to look for food. Their diet is varied and includes grass, leaves, bulbs, insects, and small animals. Sometime it even includes such large mammals as hares and young gazelles.

Of all the baboons, the hamadryas lives the farthest east—it lives in the Arabian peninsula as well as in Africa. All other baboons live only in Africa. In Ethiopia, the hamadryas baboon lives near the Red Sea. Rocky coasts, dry plains, and hilly regions make up its environment. It rarely climbs trees, but can move well in rocky areas.

Harbor Seal

Phoca vitulina

Class: Mammals

Order: Pinnipeds

Family: Earless seals

Length: up to 6 feet

Weight: 100 to 400 pounds or more

Diet: fish and squid

Number of Young: 1 or 2

Home: coastal waters off North America, Europe, and Asia

PHOTO: Gohier/NATURE, © ATP-Editions ROMBALDI MCMXCIII, English version © Grolier Inc. MCMXCIII

Harbor seals are built for life in the water. They are graceful, powerful swimmers that use their large hind limbs to propel themselves through the water, and their small front limbs to steer. Harbor seals cannot turn their hind limbs forward, making "normal" land travel impossible. On land, they travel by wiggling and hunching their bodies. Imagine yourself sitting on the floor on one side of a room, trying to get to the other side without using your arms or legs. That's how this seal gets around on land!

The harbor seal's body is covered with a coat of short, coarse hair. Under the skin is a thick layer of fat called blubber. Blubber does more than insulate the seal from the cold; it also acts as a reserve food source. When food is scarce, the seal's body burns blubber for energy.

Harbor seals spend most of their time in shallow water and on land. They are often found near the mouths of large rivers. Sometimes they wander many miles upstream. Their main enemies are sharks, killer whales, and people.

Harbor seals mate in the sea, but the female gives birth on land. She usually has one offspring, or pup, at a time. The pup weighs about 22 pounds at birth, and is about 30 inches long. Like all mammals, harbor seals nurse their young. A pup nurses for four to six weeks, until the mother sends the pup away. The young seal must then survive on its own.

Harlequin Beetle

Acrocinus longimanus

Phylum: Arthropods

Class: Insects

Order: Beetles

Family: Long-horned beetles

Length of the Body: 3 to 7 inches

Diet of the Larva: decaying wood

Method of Reproduction: egg layer

Home: Central and South America

PHOTO: Lanceau/NATURE, © ATP-Editions ROMBALDI MCMXCIV, English version © Grolier Inc. MCMXCIV

The brilliantly colored harlequin beetle is named for the handsome pattern of patchwork colors on its wing covers. It has exceptionally large eyes that wrap around the base of its antennae. The male is easily recognized by his long front legs, which double the length of his body.

Harlequin beetles seek out recently dead or dying tree branches, on which they mate and lay their eggs. From the eggs hatch larvae, or immature beetles. The larvae spend four to 12 months in the decaying tree, chewing on the wood while maturing into adult beetles. Once they reach adulthood, harlequin beetles fly to new trees, where they lay eggs, and so repeat their life cycle. Harlequin beetles mate and feed on specific types of trees, such as ficus, oleander, periwinkle, and dogbane.

Recently entomologists (scientists who study insects) have discovered a fascinating relationship between the harlequin beetle and a species of pseudoscorpion, *Cordylochernes scorioides*. This tiny relative of the scorpion uses the harlequin beetle like a taxicab. Both male and female pseudoscorpions fight for the chance to climb aboard the beetle. When the newly matured harlequin beetle flies to a new tree, its hitchhiking passengers hang on for dear life. Sometimes they attach themselves by silk threads. When the beetle arrives at its destination, the pseudoscorpion lowers itself to a suitable branch on a long silk line.

Harp Seal

Pagophilus groenlandicus

Class: Mammals
Order: Carnivores
Suborder: Seals
Family: Earless seals
Length: up to 6 1/2 feet
Weight: up to 330 pounds
Diet: mainly fish and crabs
Number of Young: 1
Home: northern Atlantic and Arctic oceans

PHOTO: © Bill Curtsinger/PHOTO RESEARCHERS, © ATP-Editions ROMBALDI MCMXCIII, English version © Grolier Inc. MCMXCIII

The sweet baby face of the newborn harp seal has captured the hearts of millions and helped inspire the environmental movement. Although harp seals are not in danger of extinction, people have strongly protested the brutal slaughter of harp-seal pups.

Harp seals give birth on ice floes near Newfoundland, Canada, and in the White Sea, north of Russia. The mothers nurse their pups for about 12 days, and then leave to go fishing. The babies remain behind, camouflaged against the snow and ice by their downy white coats. They can't join their parents in the water for about a month, when their baby fur is replaced by a dark, waterproof coat.

During this first month of life, baby harp seals are hunted by club-wielding fur trappers. The hunters have clubbed many thousands of the helpless newborn seals each year. To stop the slaughter, animal defenders began spraying dye on the pups' fur. This spoils the white coats for fur making, but saves the lives of many baby harp seals each year.

Harp seals that survive to adulthood are ready to mate when they are five or six years old. As far as scientists can tell, mated couples stay together for life. Harp seals that avoid being eaten by killer whales and sharks can live 30 years or more.

Harpy Eagle
Harpia harpyja

Class: Birds

Order: Birds of prey

Family: Vultures, buzzards, and relatives

Length: 3 feet

Wingspan: 8 feet

Weight: 10 to 15 pounds

Diet: mainly monkeys and sloths

Number of Eggs: 1 or 2

Home: Mexico south to Argentina

PHOTO: Carmichael/NHPA, © ATP-Editions ROMBALDI MCMXCII, English version © Grolier Inc. MCMXCII

The mighty harpy eagle is named after a monster from Greek mythology. The legendary harpies were horrid creatures with the heads of women and the bodies of vultures. They snatched away the food of people cursed by the gods. They kidnapped people and were thought of as demons of death.

The harpy eagle is the largest and most ferocious of all eagles. This bird of prey has extremely powerful legs, with feet as big as a person's hand. Its large bill ends in a sharp hook. It has excellent eyesight and hearing. Harpy eagles prey on monkeys, sloths, tree porcupines, and macaw parrots, grabbing these victims with wicked-looking claws that are thicker than the claws of a grizzly bear! Harpy eagles that live near people's homes and farms will often grab and devour dogs, chickens, and even baby pigs.

Harpies live on the edge of tropical rain forests, often along the shores of rivers. They build their nests, called aeries, in the tops of very tall trees. People walking through the area can often find the site of a harpy nest by looking at the ground: piles of prey bones often accumulate under the nests. The high nest protects the young eagles, or eaglets, from snakes and other predators. This is important because the eaglets are helpless for a long period of time. They are fed by their parents for at least nine months, until they are able to fly well. When they are hungry, they utter high-pitched screams and flap their wings.

Hawaiian Monk Seal
Monachus schauinslandi

Class: Mammals

Order: Carnivores

Family: Earless seals

Length: up to 7½ feet (female) up to 7 feet (male)

Weight: up to 600 pounds (female); up to 380 pounds (male)

Diet: fish and squid

Number of Young: 1

Home: Hawaiian and Laysan islands

Like its only surviving cousin, the Mediterranean monk seal, the Hawaiian species is on the brink of extinction. Although it is protected from hunters, this shy seal suffers greatly from human disturbance. Unfortunately, it must compete with tourists for the beautiful sandy shores where it lives. In the late 1980s, biologists estimated there to be only about 1,000 Hawaiian monk seals left. They believe that there are still fewer today.

So shy is the Hawaiian monk seal that it has learned to hunt primarily at night. Most likely, it chases fish and squid just below the surface of the water, not too far from the shore.

Unlike most seals, it is the female monk seal who is the larger of the two sexes. The Hawaiian monk seal is also unique in that the females do not gather in large groups to give birth. Between March and the end of May, the mother-to-be crawls out of the ocean and travels to a warm, sandy spot far from the water's edge. There she gives birth to a single pup. The newborn weighs about 35 pounds and is covered with soft black fur. The baby can swim just four days after it is born, but it will stay with its mother, nursing, for at least six weeks. During this time the devoted mother seal never leaves her pup's side—not even to find food for herself.

Heart Urchin

Echinocardium cordatum

Phylum: Echinoderms

Class: Sea urchins and sand dollars

Order: Heart urchins and sand dollars

Family: Heart urchins

Length: up to 4 inches

Diet: decaying plant and animal matter

Method of Reproduction: broods eggs inside body

Home: temperate waters worldwide

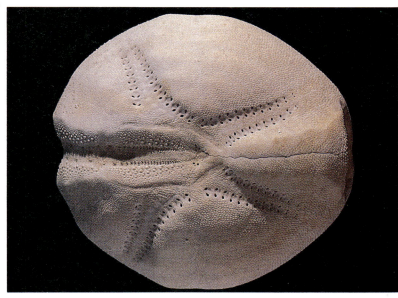

PHOTO: © G.I. Bernard/ANIMALS ANIMALS, © ATP-Editions ROMBALDI MCMXCIV, English version © Grolier Inc. MCMXCIV

As its common name suggests, this sea urchin is heart-shaped. Most people know heart urchins from their attractive skeletons, which are often sold in tourist shops. Live heart urchins are seldom seen, because they bury themselves in mud and sand.

Unlike most sea urchins, which are shaped like pincushions, the heart urchin is basically oblong. Its shape and its habit of lying buried in the soil have earned it another name: the sea potato. And while other urchins have long, dangerous spines, the heart urchin is furry. Its hairs are made of short, harmless spines.

Inside its underground burrow, the heart urchin's mouth faces forward.

Around its mouth, like the petals around a blossom, are small, rubbery feet. The heart urchin uses these tiny feet to scrape dirt and decaying matter into its mouth. Whatever it cannot digest is passed out through a hole at the back of its body. The urchin keeps a small opening at the top of its burrow through which it extends a long tuft of thin, flexible spines.

Male and female heart urchins look exactly alike. The males spray sperm into the water above the burrow. The sperm floats into the burrows of female heart urchins and fertilizes their eggs. Each female broods her eggs inside her body until they hatch.

Heath Hen

Tympanuchus cupido cupido

Class: Birds

Order: Game birds

Family: Pheasants

Subfamily: Grouses

Length: about 16 inches

Diet: leaves, fruits, grains, and insects

Number of Eggs: 10 to 12

Home: Massachusetts and the Carolinas

Date of Extinction: 1932

PHOTO: © Eric Alibert, © ATP-Editions ROMBALDI MCMXCIII, English version © Grolier Inc. MCMXCIII

In the years before the Revolutionary War, heath hens crowded the brushy plains and meadows of Massachusetts and the Carolinas. Native Americans and European settlers depended on the bird's meat.

Like the prairie chickens of today, heath hens were famous for their spectacular courtship dances. In spring, male birds gathered in large groups, jumping and running at each other with their tail and neck feathers raised. The dancing roosters produced loud booming sounds by blowing up and popping the bright-orange pouches under their neck.

By the late 1700s, New Englanders came to realize that the heath hen population was diminishing. In addition to overhunting by humans, the bird was killed by people's cats and dogs. Still more died from diseases caught from domestic poultry. The heath hen had little chance to recover, because its prairie home was rapidly replaced by farms.

By 1870 the heath hen survived only on a few New England islands. In 1907 a portion of Martha's Vineyard was set aside as a refuge for 60 remaining birds. For a short time, things looked good. By 1916 the refuge population had risen to 2,000. Tragically, a natural fire destroyed much of the hens' nesting areas in 1916, and poachers killed most of the survivors. In March 1932, the last heath hen died of old age.

Hermit Thrush
Catharus guttatus

Class: Birds

Order: Passerines

Family: Thrushes, kinglets, and gnatcatchers

Length: 6 to 7 inches

Diet: insects and fruit

Number of Eggs: 3 to 6

Home: United States and Canada

PHOTO: © Stephen Collins/PHOTO RESEARCHERS, © ATP-Editions ROMBALDI MCMXCII, English version © Grolier Inc. MCMXCII

The long, melodic call of the hermit thrush is one of the most beautiful songs produced by a North American bird. It begins with a long, flutelike tone. Then the bird repeats one melody at different pitches. The singer is always a male announcing his possession of a particular piece of territory.

Hermit thrushes usually live deep within dry pine and deciduous forests. Occasionally they inhabit swamps and household yards in rural areas. They usually nest on the ground, but sometimes make homes on a tree's lower branches. This bird makes its nest out of twigs, grass, moss, and rotten wood. The nest is then lined with pine needles and other soft bits of plants. A female hermit thrush lays her eggs in May or June and sits on them for about 12 days before they hatch. At first the mother feeds inchworms and caterpillars to her young. But after a few days, she will also give them grasshoppers, moths, beetles, and spiders. Occasionally she will even bring salamanders for the young thrushes to eat.

Hermit thrushes are very quick to capture a beetle, ant, or other bug that is crawling along the ground or on a low tree branch. By looking for food only on or near the ground, the hermit thrush avoids competing with other birds who eat the same type of food, but who find it in other places.

Herring Gull

Larus argentatus

Class: Birds

Order: Gulls, terns, and skuas

Family: Gulls

Length: 22 to 26 inches

Wingspan: 54 inches

Weight: up to 2½ pounds

Diet: fish, marine animals, eggs and young of other seabirds, and garbage

Number of Eggs: 2 to 3

Home: North America, Europe, and Asia

PHOTO: Berthon/NATURE, © ATP-Editions ROMBALDI MCMXCII, English version © Grolier Inc. MCMXCII

Of the many species of sea gulls in the world, the large white herring gull is probably the most familiar. It can be found in enormous numbers on both sides of the Atlantic Ocean. A few smaller populations spend the winter on the Pacific coast.

The herring gull's tough yellow beak has a red spot on the underside. Baby gulls will peck at this spot when they are hungry. In response, its parent will regurgitate already-digested food from its stomach for the chick to eat. Adult herring gulls are not picky eaters. Like vultures and other scavengers, the herring gull hovers high in the sky over its next meal, usually remnants thrown overboard from fishing boats. The herring gull's broad appetite sometimes takes it far away from the ocean. It often appears inland, following farmers plowing their fields. The bird eats the worms and grubs that are turned up by a tractor. And herring gulls frequently visit town dumps for a snack.

Herring gulls will often catch fish that swim close to the surface of the water. The herring gull fishes from the air. Its strong, long wings enable it to move about easily in the sky, gliding both forward and backward. When it dines on shellfish, the herring gull carries the shell high over a bed of rocks, a paved road, or a bridge, and then drops it. Its dinner will crack open when it hits the ground, and the bird is then able to feast easily on the meat inside.

Hippopotamus

Hippopotamus amphibius

Class: Mammals
Order: Even-toed hoofed animals
Family: Hippopotamuses
Length: 10 to 11½ feet
Weight: 3000 to 7000 pounds
Diet: herbivorous
Number of Young: 1
Home: tropical Africa

PHOTO: Sauer/NATURE, © ATP-Editions ROMBALDI MCMXCII, English version © Grolier Inc. MCMXCII

The hippopotamus, or "hippo," spends most of the day sleeping on river shores. To escape the strong African sun, however, it also enters the water and lies there without moving. Only its eyes and nostrils may appear above the surface, like submarine periscopes. It is also a good swimmer and diver. It can stay under water for 3 to 5 minutes with its ears and nostrils closed.

The skin of the hippopotamus produces an oily liquid that screens out the sun's drying rays. The oil also prevents the hippo's skin from being damaged by its long stays in the water. The small drops of this liquid have a reddish sheen. It was once believed that the hippopotamus was sweating blood.

In mating season, the males are sometimes seen in violent combat, each one trying to plant its large canine teeth into the body of its rival. At that time, hippos may also attack canoes, although they generally do not bother people. Mating takes place in shallow water. A young hippopotamus can only be nursed under water, but to escape from crocodiles, it may take refuge on its mother's back.

Hunted for meat, fat, leather, and ivory from its teeth, the hippopotamus has become rare in west and south Africa. Many still live along the rivers of east Africa. Its closest relative, the pygmy hippopotamus, lives in the forests of west Africa.

Hog Badger

Arctonyx collaris

Class: Mammals

Order: Carnivores

Family: Mustelids

Subfamily: Badgers

Length of the Body: 2¼ to 3½ feet

Length of the Tail: 5 to 7 inches

Weight: 15 to 31 pounds

Number of Young: 2 to 4

Diet: fruits, worms, insects, and other small animals

Home: Indochinese Peninsula and Sumatra

PHOTO: © C.B. & D.W. Frith/BRUCE COLEMAN INC., © ATP-Editions ROMBALDI MCMXCIII, English version © Grolier Inc. MCMXCIII

The hog badger is a great mystery to scientists. According to native people, this Asian badger is not rare. But it is seldom seen, for it comes out of its burrow only at night. The hog badger may also spend some of its active time hidden in the trees. It is a quick and agile climber.

The most remarkable feature of this species is its long, movable snout, which ends in a hairless nose. It is this piglike snout, along with the animal's protruding teeth, that give the hog badger its name. This creature uses its flexible nose to root around for insects, worms, and fallen fruit in the moist dirt of the tropical forest floor. The badger has a keen sense of smell and can locate its dinner in the dark.

The few people who have seen a hog badger in the wild say it can be fearsome. When cornered, the hog badger arches its back and bristles its fur to appear larger than it is. It growls and grunts angrily, and then rears up on its hind legs. In this position, it can lash out with its long, sharp teeth and front claws.

Scientists don't know anything about the hog badger's social life, how it reproduces, or what its babies look like. Of the few hog badgers kept in zoos, none have mated successfully. The oldest captive hog badger lived to be 14 years old.

Honey Buzzard

Pernis apivorus

Class: Birds

Order: Birds of prey

Family: Hawks and their relatives

Length: 20 to 24 inches

Weight: about 25 ounces

Diet: honey, insects, small animals, and eggs

Number of Eggs: 1 to 3

Home: Summer: Europe and Asia; Winter: central Africa

Summer ■ Winter

PHOTO: Lanceau/NATURE, © ATP-Editions ROMBALDI MCMXCIV, English version © Grolier Inc. MCMXCIV

The honey buzzard loves to tear apart beehives and wasp nests—not so much for the honey, but rather to eat the insects' immature larvae. The attacking honey buzzard is usually counterattacked by an enraged swarm of adult bees or wasps. Fortunately for the bird, it has extra layers of feathers to protect it from stings. The clever buzzard even catches and eats some of the attacking insects. Before swallowing, it carefully nips off the poisonous stingers.

Like other buzzards the honey species has broad, strong wings, which enable it to ride on warm, rising air currents. It can be recognized overhead by the unique way it hunches its wings while curling up their tips. Honey buzzards circle and soar above their home terri-
tories. Typically the territory is shared by a mated pair. The male, though smaller than his mate, is the territory's main defender. To announce his ownership, he performs a showy display flight, clapping together the backs of his wings.

Each mated honey-buzzard pair build a nest of sticks and leaves high in the trees. Sometimes the couple simply enlarge an abandoned crow's nest. The male and female take turns sitting in the nest, incubating the eggs for more than a month. Once the chicks are born, their parents continue to share duties. In addition to feeding the young, the adults teach them how to find and attack beehives and wasp nests.

Honeybee
Apis mellifera

Phylum: Arthropods

Class: Insects

Order: Social insects

Family: Honeybees, bumble-bees, and their relatives

Length: ³/₄ to 1 inch (drone and queen) ¹/₂ inch (worker)

Diet: pollen and nectar

Method of Reproduction: egg layer

Home: worldwide

PHOTO: Lanceau/NATURE, © ATP-Editions ROMBALDI MCMXCII, English version © Grolier Inc. MCMXCII

The honeybee may well be the most valuable of all insects. Each year, it produces millions of dollars' worth of honey and beeswax. Even more important is its work as a pollinator. As a honeybee settles on a flower to collect nectar and pollen, it transfers pollen from one flower to another, thereby fertilizing them. This leads to the formation of fruit and seeds. If honeybees were to disappear, many kinds of plants would also disappear.

Honeybees live in highly organized colonies. The colony contains one queen bee, whose job is to lay eggs. There may be up to 60,000 sterile females in a colony. Called worker bees, they gather food, build and guard the home, and care for the eggs and young bees. In the summer, there may be up to 100 male bees, or drones, in the colony. Their function is to mate with the queen. Each colony lives in a hive that contains wax combs made of six-sided cells. Some cells are used for storing pollen and honey. Other cells are used to house eggs and young bees (larvae). The queen lays as many as 2,000 eggs a day. During her four- to five-year lifetime, she produces about 2 million eggs! When a colony becomes too big, it produces a new queen. Several larvae are fed a special food called royal jelly. One of these larvae becomes a queen. The old queen and several thousand workers then leave the hive and establish a new colony.

Hooded Seal
Cystophora cristata

Class: Mammals

Order: Seals

Family: Earless seals

Length: 8 feet (male) 7 feet (female)

Weight: 880 pounds (male) 770 pounds (female)

Diet: mainly fish and squid

Number of Young: 1

Home: northern Atlantic and Arctic oceans

PHOTO: © W. Curtsinger/PHOTO RESEARCHERS, © ATP-Editions ROMBALDI MCMXCII, English version © Grolier Inc. MCMXCII

The hooded seal is named for the inflatable sac, or hood, on its nose. This sac is hollow and connected to the nose passages. It is bigger in males than in females. When a male hooded seal is excited or annoyed, he inflates the pouch with air to make loud noises.

Hooded seals have small front flippers and large rear flippers. The rear flippers are turned permanently backward. As a result, hooded seals move awkwardly on land, and spend very little time there. Most of their lives are spent in the ocean or on ice. They can stay upright in water, using their flippers to tread water. These animals feed mainly on fish such as perch, halibut, and cod; they also eat squid. Their enemies include polar bears, sharks, and humans. People kill young hooded seals for their fur.

During most of the year, hooded seals live alone. Small groups gather in early spring at the breeding grounds, where the young, called pups, are born. They are covered with a coat of dense white fur. The fur protects the pups against the freezing temperatures and camouflages them on the ice. A mother nurses her pup for only 12 days. The pup grows quickly during this nursing period because the mother's milk is very rich, consisting of about 50 percent fat. When nursing ends, the adults mate and migrate north. The pups are left to care for themselves.

GLOSSARY

algae Any of various one-celled or many-celled plantlike living things that do not have true stems, roots, or leaves, but which can make food from sunlight. Most algae live in water and are often called seaweeds. Although not true plants, algae are living organisms that resemble plants in some ways. Like plants, they can convert sunlight into starches and sugars. Many algae are too tiny to be seen without a microscope, but some algae can grow hundreds of feet long.

amphibian A class of vertebrates characterized by birth from eggs in water or a moist environment, usually followed by metamorphosis into a land animal. Unlike most reptiles, amphibians have no scales and breathe in part through their skin. Familiar amphibians include frogs, toads, salamanders, and newts.

arthropods Members of a phylum of living things that are characterized by a segmented body, hard outer skin, and jointed legs. Common arthropods include the insects, spiders, and crustaceans. Arthropods are the most numerous animals on earth.

bovines Cattle, buffalo, and bison; or, a family of animals that includes antelope, sheep, and goats.

carcass The dead body of an animal. Some animals, such as California condors, feed on the carcasses of other animals. Carcasses that serve as a food source are often called carrion.

carnivore Scientists use the word carnivore with 2 related meanings. (1) Animals whose main food source is other animals. (2) Members of an order of mammals that includes the dogs, bears, raccoons, weasels, otters, cats, and their relatives. Originally *carnivore* meant "meat eating," and is still applied to all animals that eat other animals exclusively. Common large animals that have this kind of diet include members of the dog family, the weasel family, the otter family, and the cat family.

chrysalis Another name for pupa, the resting stage that certain animals go through, especially when the pupa is covered by a protective outside coat; also, the protective outside coat. Many arthropods go through several stages before becoming adults. Insects such as ants, bees, wasps, butterflies, and moths follow a pattern in which the first, or larval, stage after hatching from the egg resembles a worm—a grub for beetles, a maggot for flies, or a caterpillar for butterflies. The second stage is a resting stage during which the insect changes its shape completely to the adult form. During this stage, a hard coating may form to protect the changing insect. This hard coating is what is meant by *chrysalis*, although the whole insect during that state may also be called a chrysalis.

cocoon A protective case for a vulnerable life stage of an arthropod, such as the egg case of an earthworm, the case for the pupa of the silkworm moth, or the case in which spiders hatch. Most arthropods go through stages where they are defenseless because they are changing from one form or size to another. In many cases, they are protected during this state by a cocoon. Spiders and many other build their cocoons from silk; indeed, our silk fiber is made from the unwound cocoons of silkworm moths.

colony A group of living things that live or grow together, such as the corals in a single formation or a hive of bees. Some coral colonies grow in regular shapes that are easily recognized. Members of this kind of colony always live together as adults but may travel alone in early life stages. The idea of colony also extends to communities of animals that just live together to survive. A bee colony is a good example. In most hives, the bees have developed a rigid social system in which each bee has a specific job. Ants and termites have similar colonies. A further extension of this idea applies to animals that can survive on their own, but choose to live in the same place, such as a beaver colony or a prairie-dog town.

endangered species A species that is in serious danger of extinction. The US Fish and Wildlife Service is responsible in the United States for listing which species are endangered. International organizations also identify endangered species. Laws in the United States and some international treaties make it illegal to hunt or trade in endangered species or to damage the environment so that it harms an endangered species. Animals or plants covered by such laws are said to be *protected*. Well-known endangered animals include the California condor, the blue whale, the ivory billed woodpecker, elephants, and the whooping crane.

environment A group of living things and the place they live taken as a whole; often, all living things on Earth and Earth as a place to live. Since people are living things, they are part of the environment. Often *environment* is used loosely to mean places where creatures live with few or no people, but houses, factories, and cities are environments as well.

evolution The process by which new species are created. Evolution occurs because animals that are better fit for their environment thrive more than less fit animals do. For example, the ancestors of giraffes lived on leaves that grew at the top of trees. The prehistoric giraffes that had the longest necks could reach the leaves, and therefore thrived, while the shorter-necked ones died out. Thus, today's giraffes all have long necks.

extinction The state of a species or other classified group (such as a family or order) when no living members remain. No species has lasted throughout all of Earth's history. Most species last a few million years. Then, usually because of changes in the environment, the last members of the species die. This is called extinction of the species. Although all species become extinct, many leave behind related species to carry on. Thus, extinction is a natural process. Today the environment is changing rapidly because of the growth of human population and new technology. Many species are becoming extinct in a short period of time. This is not good for the environment as a whole, so scientists are trying to find ways to slow down the process of extinction. Animals that are extinct are marked with a X.

habitat The place or environment where a plant or animal lives. A habitat must furnish certain basic needs for the animal. The climate must suit the animal. There must be enough food available for the animal to survive. There must also be places in which the animal can hide from its enemies and bring up its young. Some animals, such as rats and squirrels, can live in a variety of different habitats, Others, such as the koala bear and the panda, can only live in very specialized habitats. Another term often used with habitat is *niche*. The niche of an animal is the way an animal interacts with all the other creatures in its habitat. Many animals that share the same habitat have completely different niches.

heath A large, open piece of land that is not farmed and that is covered with certain low-growing plants, called heaths and heathers. Such plants are native to the Old World, so there are no true heaths in North America. Heaths and heathers grow well in soils that form where there is a lot of rain and a cool climate.

herbivore An animal that eats plants. Many herbivores are hoofed mammals called ungulates. They feed on such vegetation as grasses, twigs, leaves, and acorns. Some herbivores have a four-chambered stomach from which they bring back partially digested food (cud) for further chewing. Herbivorous mammals include horses, cattle, sheep, deer, giraffes, elephants and manatees.

hibernation The practice of some animals in temperate or cold climates of sleeping through the winter. In true hibernation, most of the body systems are shut down, and body temperature is the same as that of the environment as long is it does not get too cold. Many amphibians, reptiles, and small mammals hibernate. Larger animals, such as bears, may sleep a lot in winter and stop feeding, but their body temperature stays fairly high and they awaken from time to time, and thus are not true hibernators.

invertebrates Animals without backbones or internal skeletons. Although such familiar animals as fish, amphibians, reptiles, birds, and mammals are all vertebrates, most animals are invertebrates. Many, such as insects, shrimp, and clams, have hard outer coats instead of skeletons.

larva The stage after the egg of any animal that undergoes metamorphosis. Most commonly one thinks of the larvae of insects, such as caterpillars. But many animals change form as they grow, including sponges, jellyfish, lobsters, and frogs. For most of these the larvae (pl. of *larva*) are as different from the adults as caterpillars are from butterflies.

lichen A life form consisting of a fungus and either a blue-green alga (cyanobacterium) or a green alga that habitually live together. Lichens are often the first life forms to colonize newly exposed rock and

they are a major food source in the arctic and antarctic.

mammal A class of vertebrates that have hair and feed their young on milk produced by the mother. There are three types of mammals: Monotremes, marsupials, and placentals. Monotremes, such as the platypus and the echidnas, lay eggs which they must then incubate. Marsupials, such as kangaroos and opossums, have pouches in which the newly born young spend their first several weeks. Placental mammals, such as dogs and elephants, give birth to live young. The young usually require much care by the parents before setting out on their own. Some mammals—bats—fly, and others—whales and dolphins—live entirely in the water, but most live at least part of the time on land. Familiar mammals include pandas, horses, lions, and people.

mandible The meaning of mandible is different for different kinds of animals. Among most vertebrates—fish, amphibians, reptiles, and mammals—the mandible is a jawbone, especially the lower jawbone. For birds, however, both the upper and lower parts of the beak are called mandibles. In any grasping mouth, parts that protrude from the face are called mandibles.

marsupial An order of mammals whose young undergo most of their development outside of the body of their mothers, most often in pouches. Most marsupials are found in Australia, where they have adapted to a variety of habitats. In North America, however, the opossum has become a familiar sight. Other familiar marsupials include the kangaroo and the koala bear.

metamorphosis A change from one form to another in an animal after it is born or hatched. Although humans and other mammals change as they grow, their basic appearance and function changes gradually and not very drastically. Other animals, ranging from oysters, flies, and butterflies to frogs and salamanders, go through a period of rapid and fairly complete change called metamorphosis. For example, a frog changes from a tadpole with gills, fins, and a tail to an adult frog with lungs, legs, and no tail. Hatching from an egg is not metamorphosis since essentially the animal is the same before it hatches as after it has hatched.

migration Regular travel from one environment to another. Migration follows regular paths between 2 regions to take advantage of changes caused by the seasons. For example, many birds that live in North America over the summer migrate to warmer regions when the weather starts to become cool. Migrating birds often follow the same route each year. In a similar manner, some animals migrate down from the mountains into the warmer valleys during the winter. In tropical regions, some animals migrate to avoid dry seasons.

molting Shedding an outside coating, such as hair, feathers, skin, or an outside skeleton; also the shedding of horns by deer. Most often we think of molting among birds that lose their feathers once a year; among reptiles that shed their skin as they grow; and among such arthropods as lobsters, crabs, and spiders that shed their outside skeletons as they grow.

monotremes Mammals that lay eggs; the platypus and several echidnas are the only representatives still in existence. Scientists consider monotremes to be the most primitive form of mammals alive. In fact, monotremes and reptiles have many features in common.

nectar The sweet liquid used by plants to attract bees, butterflies and moths, hummingbirds, and other animals to the plants' flowers. Nectar is the highly nutritious reward that animals get for carrying pollen from one flower to another. Bees use it as the principal ingredient in honey.

nymph A life stage of an insect that resembles an adult but that lacks functional reproductive organs and that almost always lacks wings (even though the adult may have wings). Some insects—such as dragonflies, mayflies, and bugs—have a nymph stage, while others—such as house-flies, bees, and beetles—have larvae that do not resemble the adults.

omnivore An animal that eats both plants and other animals. Most animals are omnivorous. Even cattle eat small insects that inhabit the grass they feed on and both dogs and cats are happy to eat prepared food that includes grain or other plant matter.

plateau A fairly flat region that is higher than much of the area around it. Plateaus are often called tablelands because they stand prominently above their surroundings.

predator An animal that regularly kills other animals and eats them. Not all animals that eat other animals are predators. Some usually consume animals that are already dead while others, parasites, live off other animals that are still alive. Predators are found in all of the groups of animals, but we usually think of carnivores, birds of prey, snakes, and such insects as praying mantises as predatory animals.

prey The animals that a predator kills and eats. Prey are always animals, so grass is not the prey of cattle, but mice are the prey of cats, and deer are the prey of wolves. Small predators can later become the prey of larger predators.

pride Among lions, a pride is a group that associates and hunts together. A pride may consist of 10 to 30 lions. The pride usually includes several adults of each sex as well as 2 or 3 generations of their young.

primate An order of mammals that rely heavily on sight and have nails on their fingers and toes instead of claws. Most primates live primarily on fruit, but will eat almost anything. Two large groups of primates are the prosimians, which include tree shrews, lemurs, lorises, and so forth; and the monkeys, divided into the New World monkeys and the Old World monkeys. A small group of primates includes the great apes and people.

reptile A class of vertebrates that have scales, breathe air through lungs, and lay eggs with hard covers or give birth to live young. The long-extinct dinosaurs are considered reptiles. The common reptiles of today include the turtles and tortoises, the lizards and snakes, and the crocodilians.

rodent An order of small mammals that are characterized mainly by teeth designed for gnawing. Gnawing is a successful adaptation, for there are more rodents and more rodent species than any other order of mammal. Rodents can be grouped into three main types: those that resemble squirrels; rats and mice; and those that resemble porcupines—although usually not with quills.

scavenger An animal that feeds primarily on carrion—the bodies of dead animals. Most may also kill other animals for food. Notable scavengers include condors, vultures, and hyenas.

symbiosis Any regular close relationship between 2 species. Sometimes the relationship is helpful to both species, as the case with lichens. In other cases, it is not clear that symbiosis helps both parties; for example, the clown fish is protected by living close to sea anemones, but it is not known if the sea anemone either benefits or is harmed. Another common form of symbiosis occurs when one living thing harms the other. This is called *parasitism*; an example would be the parasite that causes malaria in humans. This parasite also has a symbiotic relationship with mosquitoes.

territory A region that an animal reserves for its own use or for the use of its own group, excluding other members of its species. Male songbirds often defend a particular territory against other males, while other animals use special scents to mark boundaries of their territory. A band of monkeys uses its territory to protect its food supply, while a pair of seabirds may use their territory to reserve a good nesting site.

vertebrates Animals that have a true backbone and a braincase. There are 7 classes of vertebrates: lampreys and hagfish; sharks and rays; bony fish; amphibians; reptiles; birds; and mammals. All vertebrates are part of a phylum named Chordata, which also includes a few animals that have spinal cords but no backbone.

wean To change the diet of a young mammal from its mother's milk to some other source of food. All mammals live on milk produced by their mother's special glands when first born, although humans sometimes substitute other forms of milk. *Wean* is also used to refer to any time that a growing animal stops depending upon its parents to obtain its food.

KINDS OF ENVIRONMENTS
Arctic and Antarctic At the top and bottom of the world are environments where the weak sunshine does little to ease the very cold weather that lasts almost all year long. Both the arctic (north pole) and antarctic (south pole) environments have ice shells

and polar ice caps, where it is always cold. Seals live in both areas. Penguins live in the antarctic, while polar bears live in the arctic. The lands just south of the icy arctic areas are called tundras. In these treeless regions, plants grow in a shallow layer of soil that thaws for a few months each summer. The tundra supports such animals as the reindeer and the caribou (actually the same species, but with different names in Eurasia and North America), ptarmigan, the arctic fox, and lemmings. Tundras do not occur in the antarctic because the limited amount of land near the south pole is covered with ice.

Cities, Towns, and Farms Cities, towns, and farms were built in places that were once forests, grasslands, or deserts. In and around human–built structures, many animals have found an environment with plenty of food and few predators. The brown rat and the house mouse prefer people's homes and farm buildings to their original environments. Pigeons flourish in cities rather than along the rocky ledges where their wild cousins live. These animals have chosen on their own to live in an environment created by people. People have chosen certain animals to tame, or domesticate. Such domestic animals include farm animals, such as cattle, sheep, pigs, horses, chickens, and honeybees, as well as pets such as the dog and cat. Although some of these animals—pigs, horses, cats—have wild populations, the great majority still live in environments created by people.

Deserts When there is too little rain for most grasses to grow, a desert environment develops. People think of deserts as hot, because the famous deserts of Africa, the Middle East, and North America are hot regions. In parts of the interior of the Sahara in Africa, for example, desert conditions are too extreme for wildlife to survive. But some deserts, such as the Gobi, can be quite cold; and the frigid ice cap of Antarctica is, in terms of precipitation, technically a desert (although it is not classed that way). Deserts look uninhabited by animals, but most are not. The reason that they appear empty is that nearly all these animals are most active at night, when it is cooler. Some desert animals live near springs or the occasional oasis, where there is more plentiful water. Common animals of the American desert are owls, rattlesnakes, pack rats, scorpions, and jackrabbits.

Forests and Mountains This category includes all forest environments except for rain forests. There are two main types of forests: coniferous and deciduous. Coniferous forests are made up of evergreen trees. They tend to grow in cool, dry climates and on mountains. Deciduous forests are made up of trees that shed their leaves each year. They grow in warmer, moister areas. In many parts of the world, forests grow that contain both types of trees. Many familiar animals live in a forest environment, including bears, raccoons, deer, and owls. Mountains generally have different wildlife than the nearby flat

areas. Usually the climate is cooler, and often there is more rainfall. Mountain environments feature sheep and goats, puma, butterflies, condors, and rattlesnakes. Animals that prefer to live on the edge of a forest instead of deep within it may spend much of their time in meadows or prairies as well as in the woods.

Grasslands and Savannas Any environment where the dominant form of plant life is grass and where there are few if any trees is called a grassland. Generally, rainfall is too little to support any sort of forest growth. Temperate grasslands occur naturally over much of the central United States and Canada and in parts of Europe and Asia. They often form the zone between desert and forest. Grasslands have also developed in once–forested areas that have been cleared for farming. Typical animals of these temperate grasslands include American bison, prairie dogs, coyote, meadowlarks, and garter snakes. Savannas are a special type of grassland that occur in hot areas with separate wet and dry seasons. The grass that grows during the short rainy season provides food for animals during the long dry season. In the famous savannas of Africa, typical animals include zebras, giraffes, lions, baboons, puff adders, kites, and termites.

Lakes, Rivers, and Their Shores Lakes and rivers, as well as brooks, creeks, ponds, and swamps, make up the freshwater environment. Although a few fish, such as salmon and some eels, spend part of their lives in fresh water and part in the oceans, most animals that live in the salt water cannot survive fresh water and vice-versa. Freshwater species include such familiar fish as black bass, bluegills, and pike, along with crayfish and some mussels. Many insects and frogs live part of their life in the water, including dragonflies, mayflies, such frogs as spring peepers and bullfrogs, and the common toad. Most of these stay near the water when they become adults, but toads move into forests or grasslands. Other animals live along the shores part of the time and in the water part of the time. These include common turtles, alligators and crocodiles, water moccasins, beavers, otters, and ducks.

Oceans and Their Shores Oceans cover more than 70 percent of the Earth's surface. Although people think of fish and sharks as typical animals of the saltwater ocean environment, oceans also teem with invertebrates of all kinds, from corals and clams to lobsters and squid. Some mammals, including whales, dolphins, and porpoises, have developed special characteristics that allow them to live out their entire lives in the oceans. Most animals live in the fairly shallow water near the shores or in offshore areas of shallower water called banks. Some animals only live in the intertidal region—the area that lies between the high-tide and the low-tide lines. A related environment occurs at the shores, the areas where the oceans meet land. Along the shores live many birds that nest near the beaches but feed on sealife. Specialized mammals such as seals breed on shores and islands, but spend much of their time hunting for fish in the water.

Rain Forests As the name implies, rain forests occur where rainfall is exceptionally high. Most familiar are the tropical rain forests found in central Africa, parts of Asia, and the Amazon region of South America. In cooler regions of heavy precipitation, temperate rain forests occur, such as the forest of the northwest coast of North America. These temperate rain forests support much wildlife, including deer, elk, flying squirrels, and many types of birds. Tropical rain forests include countless types of plants and animals, many of them unlike anything found in other environments. Tropical rain forests are famous for their monkeys, colorful birds, giant snakes, biting insects, and unusual frogs. In many parts of the world, tropical rain forests are being chopped down or burned to make way for farmland or other human uses. As the rain forests shrink, animals that live there have less space in which to live and raise families.

HOW ANIMALS ARE CLASSIFIED

Class Each phylum is made up of several classes. The mollusks, for example, include a class of snail-like creatures, a class with two shells, a class that includes octopuses and squid, plus five less familiar classes. In the phylum Chordata, which includes the vertebrates, familiar classes are the sharks and rays, bony fishes, amphibians, reptiles, birds, and mammals. Among the arthropods, the insects are considered one class, while the arachnids form another class that includes spiders, scorpions, ticks, mites, and some other creatures.

Family and genus Orders are separated into families, which in scientific classification are groups of all closely related genera (plural of *genus*). Family usually means your parents, brothers and sisters, grandparents, and other close relatives. As scientists classify animals, *family* has a similar meaning, but broader. You can recognize that a dog, a wolf, and a fox are very much alike, and indeed they are members of the dog family. Similarly, a housecat and a lion are both members of the cat family. Within a family, more closely related members are grouped into the same genus; for example, the gray wolf and the domestic dog are both in the genus *Canis*, while all foxes are in the genus *Vulpes*. Within a genus, the term *lupus* for the gray wolf and *domestica* for the domestic dog tells the species.

Kingdom Biologists generally follow a five-kingdom system, with a kingdom each for animals, plants, fungi, protists, and monerans.

Order Each class is separated into orders. The mammal class has 19 different orders, including the monotremes, the marsupials, the rodents, the primates (the order to which people belong), and the aardvark. An order is determined by common features that suggest its members are related. Orders vary greatly in their number of member species. Among the mammals, rodents—the largest order—

has 1729 species, while the smallest order, the aardvark, has but one.

Phylum There are about 30 different phyla (plural of *phylum*) of animals. Perhaps the easiest animal phylum to describe is the mollusks, even though they range from clams to giant squid. Mollusks have soft bodies, often protected by hard shells. Humans and other mammals belong to the phylum Chordata—animals with backbones.

Scientific names Scientists all over the world use the same technical name for a specific animal, called the scientific name. Scientific names are useful because the everyday common name for an animal can vary according to where it lives. One fish found from Cape Cod to Florida, for example, is called the weakfish on Cape Cod, the speckled trout in Florida, the squeteague some places in between, and a sea trout if you buy it at the fish market. But scientists all know the fish as *Cynoscion regalis*. A scientific name always consists of two parts, the genus name and the species name. Thus, any *Cynoscion* is a member of a related group (genus) of fish, but the one called *regalis* is the specific fish, or species. If three names are used to designate a specific animal, the third one refers to the subspecies. Scientists and others follow certain rules whenever they use scientific names. The names, for example, are always given in Latin, both parts of the name are printed in italics, and the first letter of the genus name is always capitalized.

Species A group of living things all considered to be representatives of the same exact type, although they may vary somewhat in unimportant characteristics, is classified together as a species. Most scientists define a species as a group of living things that regularly breed together. While this idea is often suitable, sometimes it doesn't quite hold. For example, many organisms do not breed at all; other groups that we regard as distinct species often can breed with one another. The basic idea is that a species is *one kind* of living thing, such as an Angelfish or a sea otter, as opposed to fish or otters in general.

Subspecies A population of living things that is a distinctive part of a species, often with markings or size characteristics that separate it from other members of the species, is called a subspecies, or, less commonly, a *race*. For example, all tigers belong to the same species, but the tiger of northeastern India, sometimes called the Bengal tiger, is different in size, length of fur, and color from the tiger of Siberia and northern China, called the Siberian tiger. The subspecies of Siberian tiger is larger, furrier, and paler in color than the subspecies of Bengal tiger.